Francis Banister Butts

The organization and first campaign of Battery

Fifth Series, No. 6

Francis Banister Butts

The organization and first campaign of Battery
Fifth Series, No. 6

ISBN/EAN: 9783337270971

Printed in Europe, USA, Canada, Australia, Japan

Cover: Foto ©berggeist007 / pixelio.de

More available books at **www.hansebooks.com**

PERSONAL NARRATIVES

OF EVENTS IN THE

WAR OF THE REBELLION,

BEING PAPERS READ BEFORE THE

RHODE ISLAND SOLDIERS AND SAILORS HISTORICAL SOCIETY.

FIFTH SERIES.—NO. 6.

PROVIDENCE :

PUBLISHED BY THE SOCIETY.

1896.

Francis B. Butts,
Battery E, R. I. Light Artillery
and U. S. Navy

THE

ORGANIZATION AND FIRST CAMPAIGN

OF

BATTERY E,

First Rhode Island Light Artillery.

BY

FRANCIS B. BUTTS,

[Late Corporal Battery E, First Rhode Island
Light Artillery.]

PROVIDENCE:

PUBLISHED BY THE SOCIETY.

1896.

THE ORGANIZATION AND FIRST CAM-PAIGN OF BATTERY E, FIRST RHODE ISLAND LIGHT ARTILLERY.

At the beginning of the organization of the Army of the Potomac, the commander-in-chief, Gen. George B. McClellan, was known to have said: "This is to be an artillery war." Whether this was thought necessary on account of peculiar requirements of the field of operations, or an intention of the general-in-chief to have it such by his own desire, can only be inferred. He was one of three officers sent by the war department as an attaché to the allied army in the "War of the Crimea," and by his experience at the siege of Sebastopol, and constant study, had become an accomplished artillerist and skilled military engineer. He gave special attention to the increase of this arm of the service, and from the few batteries that composed the army at Bull Run, the light artillery was increased to an average of one battery to every four regiments of infantry. It was estimated that there were upwards of two thousand guns in the several national armies, and more than three hundred in the Army of the Potomac.

Rhode Island was nobly represented in this vast power, principally through the efforts of Gov. Sprague, who had acquired great fame by his patriotic spirit, his prompt action in forwarding troops to defend the capital against the insurgents, his offer to loan the State one hundred thousand dollars to pay the expense of raising the troops before the necessary legislation could be had, and that there might be no delay in sending them forward. His gallantry in heading the troops himself, and leading them in the first battle of Bull Run, made a name for him and his State immortal in the records of the Rebellion. He had always shown strong military spirit, having attached himself, when a mere boy, to the Providence Marine Corps of Artillery, or light battery of Providence under that name. In this company he soon rose from the ranks, and was commissioned colonel by the governor of the State.

On the 1st of August, 1861, he was authorized by the Secretary of War to raise and equip a battalion of artillery, to consist of three batteries, one of which, the Second Battery, afterwards Battery A, Capt. William H. Reynolds, was then in the field. The First Battery, Capt. Charles H. Tompkins, served three months under the first call for troops, and therefore had been discharged. To this battery belongs the honor of being the first company of artillery to start to defend the capital. It was also the first battery of rifled cannon in the service of the United States.

Captain Tompkins was appointed major of the battalion, to whom the work of organization was intrusted. Battery B, Capt. Thomas F. Vaughn, was mustered into service and left Providence for Washington August 13th, and on the 25th of the same month Battery C, Capt. William B. Weeden, was mustered into service under this order, and left for the seat of war. Volunteering for the artillery had been spirited and many able-bodied men, capable of doing good service, had signed the roll intending to go. Governor Sprague asked for and obtained an order to raise and equip two more batteries to be added to the battalion. Of these Battery D was mustered into service Sept. 4, 1861, and on the 15th arrived in Washington, where Capt. John Albert Monroe assumed command. Battery E, the subject of my sketch, soon followed, D being mustered into service on the 30th of the same month, and the same day departed for Washington. On the 13th of September the war department again extended the authority of Governor Sprague, and three more batteries were ordered to be raised, the whole to constitute the First Regiment Rhode Island Light Artillery. Battery F, Capt. James Belger, and G, Capt. Charles D. Owen, followed E in quick succession, and before the last of December, 1861, all the batteries except H were in the field.

Of this regiment Major Tompkins was appointed colonel; Captain Reynolds, lieutenant-colonel; Alexander S. Webb, first major; Samuel P Sanford, second major. Colonel Tompkins resigned his commission Oct. 9, 1862, and Capt. John Albert Monroe was promoted to fill the vacancy. In June, 1862, Lieutenant-Colonel Reynolds was appointed to an important agency for the government, and Major Monroe was commissioned December 4th following to fill the vacancy. John A. Tompkins, captain of Battery A, succeeded Major Monroe. Adjt. Jeffrey Hazard, Quartermaster Charles H. Merrill, Surgeon William T. Thurston, Assistant Surgeon Francis S. Bradford, and John R. Perry, chaplain, completes the list of regimental officers as they were during the period of which I am to write. These officers, although not immediately connected with the batteries that composed the regiment, filled very important positions in connection with the artillery service, particularly Colonel Tompkins and Lieutenant-Colonel Monroe, who proved themselves to be soldiers of undaunted courage, efficient commanders, and very valuable assistants to General Barry, chief of artillery.

The organization of Battery E was begun at Camp Greene early in September, 1861. This camp was situated west of the railroad, near what is now Hill's Grove, in the town of Warwick. It was here that my experience as a soldier first began, and a visit to this locality has as much interest to me as the historic battlefields of Virginia. The camp was named in honor of Gen. Nathaniel Greene, of Revolutionary fame.

I was one of the first squad of recruits dispatched from the armory of the Marine Artillery, on Benefit Street, Providence, to Camp Greene, on the 5th of September, 1861. For several days we were the only men in camp, and having no duties except to prepare for the reception of additional recruits, we made ourselves comfortable and sought amusement in any way we pleased. Our pleasure soon ended, however, for in less than a week a number of men who expected to be officers arrived and took quarters in a tent that had been provided for their reception, and the day following we were furnished with a section of artillery (two guns and caissons) with horses. Then a squad of recruits came nearly every day, until there were about forty men all told in camp, and camp duty was begun with a somewhat military order.

This location had been selected for the rendezvous camp of the Fourth Regiment Rhode Island Infantry, by Maj. Justus I. McCarty, of the regular army, who had seen service in Mexico, and had been com-

missioned by Governor Sprague, colonel of the
Fourth, for the purpose of organizing and taking
the regiment to the field. Colonel McCarty was not
an esteemed favorite at first with the officers or men,
on account of his rigid discipline, which was thought
to be severe, at least the artillery part of the camp
thought it was, and, in order that he might reign
supreme over the whole domain, the artillery camp
was moved to the west side of Mashapaug Pond,
near Providence. Recruiting at this time was
spirited. Two batteries, the Sixth and Seventh, the
Fourth Regiment of Infantry, and the First Cavalry,
were all being organized and rapidly filling up. The
disaster at Bull Run had awakened the North to the
realization that the Rebellion was of a formidable
character, and that the South could not be conquered
without a struggle. All loyal hearts were imbued
with the highest spirit of patriotism, and troops were
rushing from every State towards the capital as fast
as they could be furnished with the arms.

Our camp was now a busy one. Two additional
guns and caissons, with horses, were furnished.
Stables were erected, guards posted, and everything

was done to give the camp a military air and appearance. We had regular drills at the manual of the piece, and occasionally a field drill with the horses. The weather was the finest of the season, and the friends of the soldiers who came to visit them were allowed perfect freedom of the camp; and, if a soldier happened to be on duty, he would be voluntarily relieved by some comrade, and the camp was one of perfect harmony and enjoyment.

The battery was fully organized in about three weeks from my first day in camp, and on the 30th of the month was paraded in a hollow square, and the oath administered binding us to serve our country for three years, unless sooner discharged. This was the last assembly that sounded for Battery E in Rhode Island and we soon left our camp. At the junction of Broad and Greenwich Streets we united with the Fourth Infantry and marched through the principal streets to Fox Point, where we embarked on board the steamer *Kill von Kull*, and took our departure amid the thundering of cannon and the mingled cheers and tears of kindred and friends. The following day we landed at Elizabethport, where

we took cars, and on the 6th of October arrived at
Washington, the seat of war, and marched immediately to Camp Sprague. The following day the
Fourth Regiment went into camp in tents near the
city, and Battery E was left sole occupant of the
camp. Those who are acquainted with the history
of the First Rhode Island Infantry, are familiar with
stories of Camp Sprague, but a brief description here
may be of interest.

Camp Sprague was named in honor of Gov.
William Sprague, who distinguished himself and
State by his active response to the calls for troops,
and personally accompanying them to the front. The
camp had an admirable location, about one mile north
of the city. On the slope of a hill, and bordering a
pleasant grove, by well laid out plans, commodious
barracks were erected for the men, each company in
line, and facing a street, at the head of which
a house, with a quaint little porch, overlooking
the camp, was the quarters of the company officers.
In the grove, sheltered from the scorching summer
sun was erected rough cottages for the field and staff
officers, and a "headquarters." These were the

2

quarters of the First Regiment. At the south, and almost uniting with the First, was Camp Clark, where the Second Rhode Island Infantry first went into camp; while at the west, and in front of the whole, was a large level space, on which was held the company and regimental drills. Our battery occupied a row of barracks at the north of the camp that had been the shelter for the First Battery, considered almost a part of the First Regiment. This camp was of the greatest interest to the members of Battery E, on account of the many anecdotes related of it by the First, which returned home July 28th, the term for which it was enrolled having expired. A larger portion of our leisure was spent in visiting the different quarters to see the decorations and emblems, some of which still remained, and to read the scribbling on the walls, which often referred to some acquaintance. A few days had elapsed when, at the hour for drill, the company was marched to the parade and Capt. George E. Randolph, promoted from Battery B, was introduced by our senior lieutenant with a few pleasant words as our commander. Captain Randolph, raising his cap, modestly said he was proud to have command of such a body of men, and hoped they would always be proud of him. How those words touched the hearts of those who heard them. Their pride for him grew to devotion. There is no officer in the whole list who became known to their men, whose name inspires such a deep feeling of endearment as the name of Capt. George E. Randolph. The officers who had accompanied the battery from the State were Lieutenants Walter O. Bartlett, William A. Arnold, John A. Perry, and Pardon S. Jastram. Lieutenant Perry was promoted to chaplain of the regiment Jan. 13, 1862, and Israel R. Sheldon was sent from home to fill the vacancy. Lieutenant Bartlett was promoted on the 24th of the same month to be captain of Battery B, and Sergt. John K. Bucklin was promoted to lieutenant. This completes the list of officers and changes during the first year of the company.

The first work of Captain Randolph was to permanently organize the company. He accordingly issued his first order dated Oct. 10, 1861, in which he announced the appointment of the non-commissioned officers, my name being first on the list of corporals.

I was then but sixteen years old, and exceedingly proud of my rank, for I felt that I must have had the confidence of my officers to have been selected from among men who were all older than myself. In this order the detachments were numbered from one to six. A detachment of artillery, in military terms, means the men that are attached to each gun, forming a body of sixteen men, including a sergeant and two corporals. The men having a desire not to be separated from those with whom they had become best acquainted, or those with whom they had together left their homes, had mutually formed detachments, which were allowed to be preserved, and the greatest satisfaction and harmony prevailed in the company on this account. Having left our guns in Rhode Island we were immediately furnished with a new battery, consisting of four ten-pounder Parrott, and two twelve-pounder, smoothbore, howitzer cannon, a complement of horses, and a full supply of camp equipage.

Our time at Camp Sprague was spent in regular drill, until November 5th, when, early in the morning, we moved from Washington and passed into

Virginia. The defenses of the capital were now being rapidly completed, and the scenes along the line of march were the most picturesque and interesting of any military operations that we had yet seen. Passing through Alexandria we went into camp about a mile southwest of the city. For a day or two we had rain, and our camp became so muddy that a more elevated spot was chosen, and we moved again, to what was known as Camp Webb. While in this camp we had our first experience of a reconnoissance. It was about three o'clock in the morning when the order came for us to hitch up. The men turned out promptly, and the battery soon moved out of camp and joined the forces near Fort Lyon, with whom we proceeded in the darkness, none of us knowing in what direction. Our march was frequently interrupted by halts, and, when daylight broke upon us we could see that we were in the midst of a large body of troops, but were entirely ignorant of the object of the movement. At noon, after marching fourteen miles, we arrived at Pohick Church, a place of historical interest, it having been the wedding place of General Washington. We saw

near the church smouldering embers of the rebel
campfires, which had the appearance of having been
hurriedly forsaken, and there not being any of the
enemy to engage our attention, the column made a
short halt, and then facing homeward resumed the
march. At this point Thompson's regulars, that
accompanied us, turned out of the road and fired a
round of blank cartridges, the object of which I
overheard him explain to our captain. He said if he
could not see the rebels he would let them hear him.
The return march was not interrupted, and we arrived
in camp about dark, having had a new experience,
that very well tested the spirit of the men, for it was
expected that we would fall in with the enemy.

After being in Camp Webb about three weeks, we
moved to the eastward, and went into camp near
Fort Lyon, where stables were erected and arrange-
ments made for passing the winter. Fort Lyon was
an extensive earthwork, on a high elevation, overlooking the Potomac and city of Alexandria, and
was one of the chain of fortications that surrounded
Washington. On the eastern slope of the hill, and
facing the Potomac near the bridge that crosses

Hunting Creek, was the artillery camp of Heintzelman's division, to which Battery E was attached.
The camp was known as Artillery Camp, and con-
sisted of Companies G, Second United States, Captain Thompson; B, First New Jersey, Captain Bean,
and the company of which I write. They were each
spoken of as Thompson's, Bean's, and Randolph's
batteries, Thompson being the senior officer and chief
of the division. Here the winter of 1861-2 was
spent in the regular duties of camp life, while the
Army of the Potomac was being organized. We
had a fine, grassy meadow for drills, in which great
interest was taken, and we perfected ourselves in
field manœuvres. Sometimes the three companies
of artillery would meet on the field together, under
command of Captain Thompson, who would take
position on a knoll, with the two junior commanders
near him, and give orders by the bugle, each battery
executing the same evolutions. These drills were
the most exciting for the men of anything that
occurred, and they would do their very best to execute the movements correctly and quickly. Battery
E so far excelled the others that these drills did not

occur as often as they possibly would, had the senior captain been encouraged by his own command. It was not unusual when we marched from the field that the Jersey battery would be seen hurrying to execute the last order, and the regulars blocked in a countermarch by having a team or two snarled up by the swing drivers, and Captain Thompson giving an exhibition of profanity that was by no means mild, even for a regular officer. Our leisure, that is when stormy weather and too much mud interrupted our drill, was spent in correspondence, reading the daily war news, and in games of various descriptions. The morals of the company at this time were of the very highest, and it was very rarely then, even for only a trifle, that any game of chance was indulged in by the men. I will say here before proceeding with my paper, that Randolph's battery was made up of men (to use a military phrase) of better calibre than any similar number of three years' men that left the State in any organization. All rules of military discipline, and orders from superiors were cheerfully obeyed. There were very few punishments, which never exceeded more than an extra hour's duty, and

with but one exception, I think, there was not a man
punished for drunkenness, during the whole period
of which I write. There was never any guard tent,
or place of confinement, nor any use for it had there
been one, and all entered into their military duties
happily and mutually.

Artillery Camp was undisturbed by any hostile
movement save one, when a call was sounded from
division headquarters that summonded us to hitch up
hastily and leave camp. It was early in the fore-
noon when we started, and as we turned into the
road leading southward, we found it filled with troops
all moving in the same direction towards the enemy's
line. We marched several miles without any indication of what was going on, when the battery wheeled
from the road and took position on a knoll that overlooked a large field on either side, and in front.
Here the battery was placed in position for action,
and a few preparations made for a defense, but, after
waiting for an hour or two without any indication of
there being an armed force near, an order came for
us to return to camp. It was then near sunset, and
without escort or company, the other troops having

gone further than we, the battery started on its homeward way. I shall never forget this as being a desolate and tedious march. The roads were muddy, and the weather about such as is experienced in the North late in November. A chilling, northwest wind, that almost amounted to a gale, numbed the men, and stiffened the mud, so as to impede our progress, and not until late at night did we arrive in camp. Fortunately I had the day previous received a box of provisions from home, sent by express, and had been well fed all day, and found my tent upright when we got into camp, while some had not provided themselves with rations, and found their tents flattened by the gale. This was our first struggle in Virginia mud, which was borne by the men with a courage and patience that must have been pleasing to the officers. A demonstration by the enemy against our picket line was the cause of the expedition.

The pleasantest memories of camp life are of our winter in Artillery Camp. The health of the men had been excellent, there having been but one case of fatal sickness, that of Corp. John B. Mathewson,

who died Jan. 7, 1862. In this case the body was
embalmed and sent to his home for burial, the
expense being borne by his comrades.

The entire Federal force had been organized by
the President and Secretary of War into three large
armies. That of the Army of Missouri and Arkan-
sas, the Army of the Tennessee or the Cumberland,
and the Army of the Potomac. The latter, commanded by Maj.-Gen. George B. McClellan, was
divided into four grand divisions, or *corps d'armée*,
commanded respectively: First, Brig.-Gen. Irvin
McDowell; Second, Brig.-Gen. Edwin V Sumner;
Third, Brig.-Gen. Samuel P Heintzelman; Fourth,
Brig.-Gen. Erasmus D. Keys. The First and Second
Corps were composed of two sub-divisions of three
brigades each, also commanded by brigadier-generals.
Each brigade was generally composed of four regiments of infantry, but in some cases there were five.
The Third and Fourth Corps had three divisions each
but were subsequently reduced to two, the same
number as the First and Second Corps. The field
artillery was distributed in about the proportion of
one six-gun battery to each brigade, and was attached

to the division the same as were the brigades, and in charge of a chief of artillery, whose next superior officer was the division commander. There was also one regiment of cavalry attached to each corps. In addition to the *corps d'armée* there was the Engineers' Brigade, composed of one regiment of cavalry, one regiment of heavy artillery, two regiments of infantry, and a battalion of United States Engineers. In addition to these there were two regiments of cavalry and two battalions of infantry attached to general headquarters, and six regiments of cavalry and twenty-three batteries of artillery in reserve. I have given this sketch of the make-up of the Army of the Potomac in order that those who were not soldiers, and some perhaps who were, may understand to what part of the army, and in what position Battery E was located, as it was attached to the First Division of the Third Corps.

The winter had passed away. The armies of the West had been pushing forward and down the Mississippi, sweeping the rebel army from before them and carrying victory wherever they went. Forts Henry and Donelson, together with Island No. 10, and the cities of Nashville and Columbus had been taken, and our army lodged in the heart of Tennessee. Nothing had yet been done to wipe out the disgrace of Bull Run, and clamors had become loud for the Army of the Potomac to move. The confidence of the soldiers and all loyal people was bestowed upon General McClellan with firm devotion, who, it was believed, had organized an army superior to any the world had ever seen, certainly superior to that of the Confederates, and that he would lead it to a speedy crushing out of the Rebellion.

At length indications gave promise that something would be done. Shelter tents were issued and preparations for the evacuation of our winter huts were commenced. The First Corps, General McDowell, had moved upon the winter quarters of the Confederates at Manassas, which were found evacuated, and all stores and army supplies removed, which indicated that they intended to give up that part of the field and make a stand nearer Richmond. A few days later everything was packed in the battery for a campaign, and on the morning of the 17th of March the battery wheeled out of Artillery Camp, leaving a

3

scene of the happiest recollections. We had a full complement of men, a battery fully equipped, and every heart throbbed with patriotism and a determination to serve the guns to the last man. Before noon we had placed our battery and horses, with the other artillery of the division, on board barges at Alexandria, and immediately commenced descending the Potomac in tow of transports that were ladened with troops. This movement could not have been a secret even to the enemy, for I remember of its being generally understood that the advance upon Richmond was to be made by way of Fortress Monroe, and the rebel iron-clad *Merrimac* having been defeated by the Ericsson battery *Monitor*, it was thought she might have been so disabled that the army would ascend to some point on the James.

The first few hours of our sail down the Potomac were delightful. As we passed Fort Washington we saw some shots fired in practice that bounded over the surface of the water or plunged beneath it, when, sometimes, if a shell, it would explode, throwing volumes of water to a great height, showing the power of heavy guns. A little further on we passed in sight of Mt. Vernon, the home of our first president. The scene was sublime. Fertile lands and pastures blooming with the first tokens of spring, and woodlands bearing the ever green foliage of southern pine, bordered the shores on either side, while the water, stretching seaward to the edge of the horizon, lay placid and mirror-like under a most brilliant descending sun, seeming to reflect upon us the peace which God had given, and to forbid our hostile intent. Later in the afternoon one of the transports in convoy separated the tow by taking two of the three barges, and proceeding left the one on which were most of our men in tow of a steamer that would have had a load to move herself, and was so slow that we all got tired out watching to pass a given point, and after disposing of some hot coffee that had been prepared for us, and an allowance of hardbread from our haversacks, we spread our blankets on deck where we slept till morning.

When we awoke, we found the barge at anchor several hundred yards from shore at the entrance of St. Mary's River, on the east side of the Potomac, near its mouth, and that the steamer had departed, leaving us with no other company than a few small oyster craft. At first we could not understand the object of our having been left unprotected in what seemed a desolate region. But as the day continued, we saw that the captain of the steamer had known by his nautical experience that a storm was approaching, and that disaster would follow if he advanced upon the broad waters of the Chesapeake. The upper deck of our transport had been fitted up with some accommodations for passengers, having a large saloon, a few rooms, and a cabin, the latter being in possession of our officers. The day was spent in idleness, and with little anxiety except that our rations were getting short, from which we were relieved by the arrival of a steamer ladened with army stores, which had been driven in with several other craft for a harbor, the storm now making it unsafe for any but large vessels to proceed. During the evening we were enlivened by the music of a troop of slaves, who visited us from the plantations. They had a banjo and violin, which were strung up into a somewhat harsh tone, as an accompaniment to male voices. They also disposed of a quantity of eatables, a description of which would reflect upon the honesty of the negroes, for there must have been a disappointment in both pantry and hen roost on more than one plantation, with the proceeds of which and a collection for the entertainment, the visitors departed happy and wealthy.

The following day everybody stayed on board the transport till evening, when Captain Randolph, Lieutenant Arnold, and Lieutenant Butler, of the Regulars, a part of whose officers and men were also on the barge, went in a boat up the river to a planter's house, where they spent an hour or more enjoying the hospitality of the gentleman and his family. A good supper was spread, of which the officers could not have partaken with more relish than one of the two men who accompanied them, who made up for lost meals and put in an allowance for future possibilities. The morning of the fourth day opened clear, and a steamer took our craft in tow at an early hour and proceeded with good speed towards Fort Monroe, where we disembarked in the afternoon. The battery and horses with which we had parted company on our passage, had arrived previously, and had been placed in waiting order by Sergt. William Millen and a few men who were with him. Without delay the battery was hitched up, and, moving through the burned village of Hampton, whose bare chimneys stood as solemn evidence of rebel vandalism, we went into camp about a mile west, near what is now the Soldiers' Home.

For several days the large fleet of steamers that it took to transport the army continued to discharge troops and war material, until the whole Army of the Potomac was in one camp, that could be overlooked from a single point. This brought us near the Second Regiment and Batteries A, B, C, and G, and I enjoyed greeting old friends, some of whom I had not seen since they left Providence. The enthusiasm at this time was intense, as it was thought nothing could resist the power of our army, and that the army of McClellan would sweep like a tidal wave over the whole seceding states.

On the morning of the 4th of April five days' rations were stowed into our haversacks, and at eleven o'clock A. M. we started with the advance divisions of Hamilton and Porter, and, marching through Big

Bethel, one of the outposts of Yorktown, made memorable by the fatal fight June 10, 1861, we arrived at Harwood's Mills, where we encamped for the night. Next morning we resumed the march, and about three o'clock reached Yorktown and immediately went into position in front of the enemy's fortications at Winn's Mill, about one mile to the left of the principal defences.

The First and Second sections of the battery went into position in front of the rebel works at Winn's Mill, with which they immediately became hotly engaged. At the same time the third section, in charge of Lieutenant Bucklyn, advanced across an open field and went into battery near a fence, along the sides of which a thick growth of brush hid our movements. A few shots, however, brought us to the notice of the enemy, who returned the compliment, making some close hits. From this position I saw plainly the effect of the firing, which afforded considerable merriment on both sides. Once I directed a shell at a company of infantry, which was seen to explode directly among them, and caused such a sudden breaking of ranks, and skedaddling, as to create a cheer from all on our side. The first two sections had taken what seemed to be a trap set by the enemy, for he had such complete range of the position that every shot took effect, and, before we got through, eight dead horses and one wheel demolished was the cost of the experiment. In return for our cheers a Confederate band mounted the parapet of their works and commenced a march up and down to the chorus of Dixie Land. We were not separated too far to hear words of command, and the sudden stop of the music and the disappearance of the band at the flash of our cannon, made fun for us all. These provoking performances were exchanged at intervals until darkness came, and we retired out of range and bivouacked for the night.

Battery C and Griffin's regular battery had been engaged since noon in front of Fort Magruder, on the Yorktown road. The Third and Fifth Massachusetts batteries were also in action on our right, whose fire was returned with spirit and determination by the enemy, causing a cannonade that seemed to us at that time terrific. Battery C lost one man killed. The Third and Fifth Massachusetts had two men killed and several wounded ; also three horses killed. Berdan's sharpshooters had done good execution with their telescope rifles, and succeeded in silencing, for a time, some of the rebel guns by picking off the cannoniers whenever they attempted to reload. These men also suffered a loss of several of their number from not having sufficient protection for use of muzzle-loading guns, there not being at that time any breech-loading military rifles in use except, perhaps, for cavalry, and might not have been now, had they not been one of the developments of the war.

The second day at Yorktown was cloudy, with some rain. Our camp was in a field not far from the works, and the general belief was that the army would be consolidated at once and an attack made. This idea was, however, given up, as there was no indication of additional troops, but late in the afternoon a section of our battery moved to the front and exchanged a few shots with the enemy, our object being to break up working parties that were busily engaged strengthening their fortifications. At night we returned to our encampment to find that it had been transferred to a reserve picket post, all the bat-

tery, except the guns and horses, having been moved
to the rear. No fires nor lights were allowed at
night, and the greatest precaution was taken not to
attract the attention of the enemy. This was one of
the most unpleasant nights I ever spent in camp.
Our five days' rations were gone, and we had been
on short allowance for two days. The roads from
Fort Monroe had become impassable owing to the
past rains. Ammunition, provisions, and forage had
to be left behind, and a new base of supplies was
looked for nearer Yorktown. Not a murmur of com-
plaint was heard from the men, who bore their hun-
ger with courage ; they had discovered that the field
on which we were encamped had been recently planted
with peanuts, which they dug up and ate. A quantity was also found in a building on the farm, and a
small cow was captured, which also furnished us with
food until the wagon train arrived.

On the morning of the third day we again went
into position in front of the rebel fortifications, which
had been considerably strengthened, and extended
in a long line till lost sight of in the woods. Our
first shot was at a party of shovelers, who dis-

appeared like magic at the flash of the gun. Not anticipating any danger except from the rebel shell, which we had learned to skilfully dodge, I mounted the top rail of a fence to see the effect of our shot, when zip-whiz, a sharpshooter's bullet came so near that I seemed to have lost half my nose, and I almost broke my neck by an impulsive twitch backward. The artist who had taken my head for a target was seen in a tree-top about three hundred yards distant on our right. He was brought to grief by Sergeant Milne, who dropped a shell exactly into his nest, and we had no further trouble from that source.

Active operations against Yorktown by this time were given up. General McClellan had decided upon the more tedious plan of digging the enemy out, so we dropped our guns and took up spades. But we did not do much digging; in fact we did nothing. If we had, the one hundred and ten thousand men that composed the army then, could have demolished the rebel fortifications with their artillery or shoveled the whole town into the ocean.

Battery E was moved to a camp in the woods about a mile farther in the rear. Our anticipations had failed, and we awaited developments. The first mail from the north brought papers giving accounts of the first assault on the rebel works, topographical plans and descriptions of their impregnability, as seen from the office in New York, where the engineer was at work. The entire rebel army was said to be concentrated here. Yorktown was to be the decisive point—"the Key to Richmond," "the Sebastopol of America." The method of conducting the war was to be changed. No more bloody encounters upon open fields, to meet face to face and man to man. This was to be abandoned, and engineering skill was to conquer. This was the tone of the daily news at that period of the conflict. The work of digging the enemy out was begun ; every soldier was armed with a shovel or pick, massive earthworks were erected in front of the rebel fortifications, extending from the York to the James River, and heavy siege guns and mortars were placed in position. Some of our infantry, holding the advance lines, would daily creep close to the rebel intrenchments to take a shot at some of the rebel artillerymen, which would often bring on, for a few minutes, a fierce cannonade.

On the morning of the 16th it was discovered that the enemy had erected a work during the previous night on the opposite side of Warwick River, between Winn's and Gee's Mills, which two Vermont regiments and three batteries of artillery were sent to attack. The Vermonters, after a brief skirmish, rushed into the stream waist deep and stormed the works. The first line was carried; but a shower of lead and iron checked the advance, and having had no support they were forced to recross the stream, suffering a severe loss. This attack brought out most of the troops along the whole line, in fear of a general engagement, and a sharp cannonade and picket-fire were kept up for some time, in which Battery E participated.

General McClellan made every effort to avoid an exchange of shots, and to restrain picket-firing. Perhaps this sport may have been indulged in more than was necessary, but our soldiers could not be kept quiet in sight of the rebels. Alternately a section of our battery would go into the intrenchments at night, to return at daylight, the rest of the time being spent in idleness within the woods, out of sight

4

and sound of what was being done. Two or three
times at this period I went along our lines and saw
with a great deal of interest, the erection of siege
batteries and the mounting of heavy rifled guns and
mortars. But this was too slow, every man wanted
to do something. Our inactivity at length brought
on discontent and strong feeling against continuing
the siege. "On to Richmond!" was shouted by
press and people until the fourth day of May, when
our impatience was relieved by the intelligence that
Yorktown was evacuated, and an order to hitch up
and be ready to follow in pursuit of the fleeing
rebels.

Three days' rations were issued and Battery E
moved out of camp, leaving behind everything not of
use in a fight. On arriving at the Yorktown road
we joined a column of troops marching forward, and
entered the intrenchments in front of the town.
Here a halt was made in consequence of the large
number of explosive shells and bombs that were
placed in the road, or wherever it was thought our
troops would pass. These were laid in all sorts of
ingenious forms for destroying the lives of our men,

showing clearly the barbarous spirit of Southern soldiers, and we saw a body of prisoners being forced at the point of the bayonet to remove the shells or mark out the spot where these infernal machines were buried. Hooker, who commanded the Second Division of the Third Corps, had entered the works about eleven o'clock, giving the rebels full six hours the start. General Hamilton had been relieved from the Third Division, to which, as I have said, Battery E was attached, and Gen. Philip Kearny had taken command, who pressed hard upon Hooker in pursuit of the retreating enemy. The advance was slow, as we had not proceeded more than three miles when darkness concluded operations for the day, and we wheeled out of the road and without shelter bivouacked for the night. Before morning the sky clouded and a drenching rain set in, giving a discouraging promise of a successful chase.

At daybreak reveille was sounded through the whole encampment, and little streams of smoke were seen in every direction, curling upwards from small fires, around which could be seen squads of soldiers busily engaged in cooking their meals. Soon the bugles

sounded " Boots and Saddles," then drums beat " Fall
in," and, turning from the field, we were again on the
road leading westward. Our progress was slow, often
waiting for more than an hour for the troops in advance to move on. Governor Sprague having joined
the staff of General Barry, chief of artillery, was
frequently seen during the day. On one of our
halts he came through the battery leaving some pleasant remark in return for the salutes which we all
felt proud to give him. We also learned from him
that Hooker had caught up with the enemy, and was
having a brisk engagement. At noon we began to
move more steadily, a part of the troops having taken
the road leading to Warwick Court House, that now
formed a junction with the Williamsburg road. The
road soon gave evidence that a large force had passed
before us. The rain of the previous night, and
tramp of cavalry and artillery had put the roads in terrible condition. The troops on foot marched through
the woods on either side of the road, while our
battery strode through mud in which the horses sank
to their bodies. Heavy firing was heard in front,
indicating that the retreating army had been over-

taken, and a battle was raging. Every means was taken to hasten the troops forward. Our guns would now and then lodge in mud-holes till it was found necessary to abandon the caissons, using the horses, making twelve on each gun, and to lighten the ammunition chests the cannoneers and infantry advancing with us carried the ammunition in their hands. As we came nearer the firing was more distinct, and every effort was made to get forward in order to reinforce the troops engaged. We were often checked by the miring of some of the guns, until we reached a road that had been chopped through the woods. A short distance further on wounded men were seen coming to the rear. A steady cannonade and musket firing was heard only a short distance in our front. General Heintzelman was then urging his men forward with loud words of command. Turning to the left we emerged from the woods, when a thick, sulphurous smoke enveloped us, and loud cheers given by Hooker's exhausted division, told of their joy when Kearny dashed upon the field. Battery E was not ordered into the fight, as it was soon over. The rebels were driven from their rifle-pits ; darkness,

and a heavy rain that it seemed would deluge the
earth, closed the scene, and the battle of Williamsburg was ended.

Halting thus upon the edge of the battlefield,
our horses were unhitched from the guns; the men,
soaked with rain, hungry, and overcome with fatigue,
laid down in the mire and slept. It was about midnight when I awoke, the rain had ceased, the sky
was unclouded, and the moon, that never seemed
brighter, cast upon us a most brilliant and welcome
light. Arising from the bed I had made of a few
fence rails, I went to a fire that I saw burning,
around which sat a dozen or more men, some wounded,
that belonged to Gibson's Battery, Third United
States Artillery. This battery went into battle in
echelon, on the main road, in front of Fort Magruder. It had become mired, was charged upon and
captured by the enemy; was retaken, but it had to
be abandoned, the guns having settled into the soft
mud, so as to become immovable. I listened to their
stories of the fight with intense interest, at the same
time kept turning myself before the fire until my
garments were quite dry.

With the first streak of daylight Captain Randolph came to where we were standing, then walked through the battery, apparently anticipating an early order to move. The men soon got up from their muddy beds and commenced making fires by which to dry their clothing. Our haversacks having been fastened to the foot-boards of the gun carriages, coffee was made, which, with a few hard crackers, made a breakfast that refreshed and strengthened us all, and at sunrise Battery E was in harness, ready to move. It was expected the enemy would continue their retreat during the night; in fact our advance pickets who lay all night very close to Fort Magruder, heard them withdrawing their artillery, and as soon as it was light went forward and took possession, so we were not suprised to find them gone. About an hour after sunrise the left section (two guns of Battery E, and a section of Thompson's battery) moved forward. A short halt was made as we crossed the battlefield, which everywhere gave proof of the terrible struggle that had taken place. The dead and wounded lay thickly over the ground, disabled cannon and artillery horses blocked the road we were to enter, and the field on the right and in front, extending to the rebel rifle-pits, was strewn with the dead of both armies, while guns and accoutrements in great quantities were scattered in all directions. Proceeding as far as Fort Magruder, we joined an infantry force under General Jameson, which immediately deployed, advancing as skirmishers to the right and left of Williamsburg, while the artillery, with cannoneers dismounted, passed through the town by the main street and halted near the ancient seat of learning, William and Mary's College, the infantry continuing across the plain entered the woods a mile or more distant and disappeared. The Eighth Illinois Cavalry immediately came forward and followed in pursuit of the fleeing rebels, who in their retreat had abandoned one large gun in the works, a light brass gun of ancient manufacture, and two caissons, besides strewing the road with muskets, bayonets, knapsacks, and clothing in great quantities.

Our halt lasted several days, and the battery, that had been scattered the whole length of the route from Yorktown, was reunited. General Stoneman, with the Eighth Illinois and Sixth Regular Cavalry,

the Second Rhode Island and Ninety-eighth Pennsylvania Infantry, and a battery of horse artillery,
had gone in pursuit of the enemy. On the 8th of
the month, three days after the battle, the army
began to move from Williamsburg, but not till the
10th did we resume our march towards Richmond.
Three days after the battle of Williamsburg General
Franklin, who had arrived with his division of reinforcements, some ten thousand men, advanced up the
York River and landed at West Point, with considerable opposition, and a severe fight occurred, resulting
in another flight of the rebels and a loss of about fifty
men killed and one hundred and fifty wounded on
our side. Porter's Division was transferred from
Yorktown to West Point during the next two days,
and on the 10th marched to White House Landing,
on the Pamunky, twenty-five miles from Richmond,
where the headquarters of the army were established
on the 16th. Heintzelman and Sumner advanced
slowly towards the new base of supplies, where the
army was being concentrated, arriving there on the
13th. On the 14th and 15th it rained and we
remained in our camp. About this time the Fifth

and Sixth army corps were organized. The Fifth consisting of Porter's division of the Third Corps, Sykes's division of regulars, and the reserve artillery, under command of Gen. Fitz John Porter. The Sixth, consisting of Franklin's division of reinforcements, and Smith's division, detached from Keyes's corps and commanded by General Franklin.

In moving towards Richmond the right of the army, Franklin's and Porter's corps, had kept to the north, striking the Chickahominy at New Bridge. The left wing, Keyes's and Heintzelman's corps keeping south, crossed the river at Bottom's bridge, while Sumner connected the right with the left. By the 28th Keyes's and Heintzelman's corps had crossed the Chickahominy, and the corps of Sumner, Franklin and Porter were stretched for eighteen miles along the opposite side of the river with headquarters at New Market, in rear of the right wing. While the army is thus extended along the region of the Chickahominy let us take a glance at the stream and its swampy borders, thousands of loyal and cherished patriots were swept from our ranks by its fevers and diseases. The river rises northwest of Richmond, where several small tributaries, flowing through swampy uplands unite, then, running in a southwesterly direction, between the York and James, and parallel thereto for about fifty miles, turns to the south; and, after a winding course of twenty miles, falls into the James ten miles west of Williamsburg, being navigable for small craft for about twenty miles. The stream is crossed by several bridges, principally those used in the military operations of the Peninsular campaign. Bottom's bridge, on the south, crossed by the Williamsburg road, and Meadow bridge, fifteen miles to the north, crossed by the Fredericksburg Railroad, are the principal of these. Richmond lies six miles from the Chickahominy, opposite the centre of our military operations. The movements of the army had been so slow in advancing up the peninsula that a well defined description of the approaches to Richmond had been pictured on the mind of every soldier who could read our daily newspapers, and in crossing the Chickahominy more than ordinary notice was taken of its formation.

On the 29th of May, Battery E crossed the Chickahominy at Bottom's bridge. In approaching the bridge a gentle incline from the bordering uplands brought us upon a broad intervale intersected by deep ditches, and now and then slight depressions, filled with stagnant water, over which logs had been laid in corduroy in order that the artillery and cavalry might pass. Further on we entered a wooded belt, where slimy water and spongy soil indicated that we were in the midst of the great Chickahominy swamp. Here we crossed Bottom's bridge, built with three short spans, and supported on piers of logs cobbed together. The work had been done by our own engineers, who rebuilt the bridge, the enemy having destroyed the one on which they crossed in their retreat. Under this bridge flowed a sluggish stream about a dozen yards wide and four feet deep, which filled its banks to the brim. This was the historic Chickahominy. The woods are about four hundred yards in width, and as we pass out on the opposite side we again cross the bottom lands and ascend to higher ground. Turning our heads to glance backwards, we see across the swamp a mile or more distant, a long column of troops, winding over a vast plain, resembling a huge serpent, until it disappears into the forest, the top of which was about on a level with the ground on which we stood. Without a turn we kept along the Williamsburg pike for about a mile, when we left the road at the edge of a wood and went into camp.

Keyes's corps had crossed Bottom's bridge two days before and occupied the position now taken by Heintzelman. Slight defences had been made, consisting of a redan on either side of the broad turnpike, and a chain of rifle-pits. In front and on either side of our camp was level ground for a distance of about a thousand yards, the opposite side being bordered with heavy woods. The Richmond and York River Railroad was less than a mile at our right. Keyes had advanced his line about a mile and was intrenched in the woods on the Williamsburg road, at "The Seven Pines," a place so named from a clump of pine trees at the crossing of several roads. Casey's division of the corps was a half mile further advanced to what is known as "Fair Oaks Farm," where there were a couple of houses in a pleasant grove of white oaks. The white oak of New Eng-

5

land, with its smooth, silvery bark, becomes scarce in traveling south, until it finally disappears, and in this part of Virginia good specimens are uncommon, and they are called fair oaks, which gave origin to the name of this famed locality. Fair Oaks Station is often confused with Fair Oaks Farm. The former is situated a half mile further north, being a wood station on the Richmond and York River Railroad. Heintzelman's corps lay behind that of Keyes, extending from the railroad, and to the left in the direction of White Oak Swamp, both corps numbering about thirty thousand men.

Our first night on the south side of the Chickahominy was marked for its pleasantness. The temperature was mild and beautiful, and the grove in which we bivouacked was filled with fragrance from the tulip trees aud flowering shrubs, while in the tree-tops, then in full summer foliage, sat numerous whip-poor-wills, striving to hush a rival or charm a mate with their enchanting songs. The moon, too, seemed to have more than its usual splendor, lighting up the scene with its silvery rays till all about us was discernible. The army was at rest. Nothing disturbed

the quietness save the muffled tramp of weary senti-
nels till daylight aroused the camp, when the busy
work of army life was renewed. No reveille is
sounded, no bugle or drum is heard, for the enemy
is near. The drivers quietly groom and feed their
horses, then from their haversacks a breakfast is prepared by the men. No dainties are had. A rough
piece of boiled salt beef, or a piece of raw salt pork,
well sprinkled with the mud or dust of the march, a
canteen of water poured into the empty tomato can,
a little coffee is added, and, proceeding to a fire that
some comrade has kindled, the water is boiled, then
from a pocket is taken a knife, having folded within
its handle a fork and spoon, and, with a few hard
crackers, this is all there is to a soldier's meal while
on a march.

The sun had scarcely risen when clouds overspread
us, and at noon rain commenced to fall. The army
was without shelter, except the miniature tent that
each man folded in his blanket. With these we
found refuge from the torrents that put an end to all
military work, filling the rifle-pits and rendering the
roads impassable. From noon till midnight the storm

was terrible, surpassing that of Williamsburg. I never before heard such thunder, and the heavens flashed with a perpetual blaze of lightning.

The next morning the sky was still clouded, the earth was filled with water, and work on intrench-ments, or the movement of troops, seemed to be impossible. Nothing of record transpired till about noon, when firing was heard on the line in front of Seven Pines. It suddenly increased; soon the sound of artillery, then the roar of battle. Kearny's division was hurried forward and passed from our sight into the woods, towards the battle. The left section of Battery E moved into the redan on the right of the turnpike, while the other two of rifled guns occupied similar works on the opposite side. The roll of musketry and boom of cannon were earn-estly listened to, as we could distinguish our own from that of the enemy, and we knew our men were outnumbered. Our hearts were firm in the courage of our men, but the position of our army, being divided as it was by the flooded Chickahominy, caused grave fears to those who understood the situ-ation. Nearer and nearer the sound of the battle came. Then fleeing troops that had been over-whelmed and driven from the field swarmed from the woods, and continued past our line of reserve in lively retreat towards Bottom's bridge. Heintzelman, with Kearny's and Birney's division, placed himself between the retreating men and their conquerors, where he stood firm as a rock. Casey succeeded in checking the retreat, and, rallying his men without any regard to formation, re-entered the fight. The vast horde that endeavored to sweep the left wing of our army was held in check. Right, left, and centre were successively attacked, and the battle raged till dark, but the enemy could make no further progress. The excitement in rear of the line of battle was at times intense. In our slight fortification was the headquarters of our corps commander, where hurry-ing aide-de-camps and orderlies swarmed with reports from the fight, and flew off on their steeds to deliver orders. The council of generals ; the hurried move-ments of troops ; the loud commands ·of officers; the buzz of an over-reaching bullet, and the flash of a musket of some far-off picket ; then the heart beats a moment in fear that our men are being driven back,

and is again rejoiced as we hear the cheers of our
own men, and the rebel yells die out. These are inci-
dents of the scene in rear of the battle.

Darkness ended the battle of Seven Pines, and
Keyes and Heintzelman consolidated their forces on
the plain in their rear, leaving a strong line of pickets on the edge of the field, which was the camping-
ground of the partially victorious rebels. The roads
were still filled with mud, and no troops could be
moved except infantry. The Chickahominy was
swollen, overflowing the swamp and bottom lands ;
the roads were flooded, and bridges washed away, so
that no reinforcements could reach us. The whole
rebel army was within sound of a gun, watching for
day to come in order to renew the attack, in hope of
vanquishing the divided Union army. Our men lay
down for rest beside their arms, with the glare of
the enemy's campfires illuminating the sky above
their miry beds. In a short time all was still, nothing save an accidental alarm disturbed their slumbers
till after midnight, when they were awakened and
marched off towards the right of the field.

At daylight all was activity at the headquarters of

Heintzelman, Keyes, and Sumner. Sumner had constructed two bridges in front of his line across the Chickahominy, by which he had crossed before they were deluged with two divisions, those of Richardson and Sedgwick, both of which had been engaged the day before at Fair Oaks Station, this being the right of the line of battle. At seven o'clock the enemy began the attack by advancing down the railroad, leaving Sedgwick, who was still at Fair Oaks Station, and fell upon Richardson's division, supported by artillery, who repulsed the attack at every point. Hooker, who had been on the left, rushed his division to the point of the heaviest firing, and, with Birney's brigade, of Kearny's division, fell upon the enemy's rear, and, after an hour's hard fighting the enemy broke and fled, taking the Williamsburg road towards Richmond in their retreat.

The battle of Fair Oaks lasted only four hours, during which not less than seventy-five thousand men were actually engaged, the Confederates greatly outnumbering the Federals. Battery E was not called into action as the battle was fought entirely within the woods, and no artillery was used on either side except a few shots from Sedgwick, when the attack was first made. The roar of musketry was terrific, more terrible than the thunder of the previous storm. Both armies fought desperately, and their commanders led them on with more determination than had ever before been displayed, both seeming to realize the very critical position they were in. One having behind it a swollen stream, and cut off from all support or retreat; the other, fighting under the very walls of its Capital, in front of its gates. After the battle I went, with a comrade, over the field, which everywhere gave evidence of the desperate encounter. Within the space of a mile lay more than seven thousand dead and wounded of both sides. The sight was appalling.

General McClellan entered the redoubt our battery was in, after the battle was over, accompanied by a few of his staff, among whom was Count de Joinville, who approached me very politely, as I stood near a gun, and asked several questions concerning the elevation and time of fuze; meanwhile McClellan, Heintzelman, Kearny, and Hooker were in conversation about the battle. This was the first appearance of the commander-in-chief, he having been unable to get across the river till that time.

After the battle of Fair Oaks, General Heintzelman made a reconnoissance, going beyond the woods considerably until within shelling distance of Richmond, and made an extended tour on both sides of the Williamsburg road without seeing any of the enemy. General McClellan did not think it would be expedient to follow up the fleeing and demoralized rebels, or to advance his line beyond the woody belt which had sheltered the Confederates. Heintzelman posted himself in the position held by Casey at Seven Pines, his right extending to Fair Oaks Station, where he joined Sumner, whose forces stretched eastwardly down to the Chickahominy. Keyes united with Heintzelman's left, the line running southerly till it reached White Oak Swamp, the shortest distance between the two wings being about three miles, and five miles around the arc. The woods were cut down, and earthworks thrown up the whole length of this line. There were several redoubts built of timber and earth, into which the field artillery was placed, Battery E occupying the

one on the extreme left of Heintzelman's corps,
where the battle began on May 31st.

About this time, I cannot give the exact date,
General Kearny issued an order that all the officers
and men of his division should wear conspicuously
on the front of their caps a square piece of red cloth
to designate them. This became known as the
Kearny patch, and other corps, one by one, adopted
distinctive badges, each division being indicated by
one of the national colors, red, white, or blue, and
green where there were more than three divisions.
These badges were used on headquarters' flags instead
of numbers, as had been used, and painted on all
the wagons, and stenciled on all government prop-
erty belonging to the corps. The soldiers were soon
supplied with metal badges, by salesmen who went
about the camps. These were made with enameled
colors, with the name of the soldier, his company
and regiment lettered upon it, and quite often a list
of his battles. Every soldier was proud of his
badge; not so much for its being an ornament, but
proud of his corps, proud of the battles he had been
in, proud to be recognized wherever he was seen.

Many a soldier, found dead on the battlefield, received careful burial, and his grave was made known by the badge found on his coat. A story is told of one of our brigades that before going into battle at Cold Harbor the soldiers pinned papers on their backs, on which were written their names and regiments. Such patriotism and heroism ought never to be forgotten. Corps badges have now a legal recognition in the Revised Statutes of the United States:

"Section 1,227 All persons who have served as officers, non-commissioned officers, privates, or other enlisted men, in the regular army, volunteer, or military forces of the United States, during the War of the Rebellion, and have been honorably discharged from the service, or still remain in the same, shall be entitled to wear, on all occasions of ceremony the distinctive army badges ordered for or adopted by the army corps and division, respectively, in which they served."

To Maj.-Gen. Philip Kearny belongs the credit of being the first to adopt, and of first having issued orders for all officers and men of his division to wear badges.

On the 13th, two weeks after the battle of Fair Oaks, McClellan moved his headquarters across the Chickahominy, establishing himself near Savage Station. Franklin's and Porter's corps, with McCall's division of reinforcements of Pennsylvania Reserves, some ten thousand men, were left on the north side, their line extending east to Cold Harbor, and having the railroad to guard by which was forwarded the supplies. In this position the entire line extended eighteen miles, leaving the right wing separated from the main body of the army, thereby dividing it by the Chickahominy, in the same manner as it did when the Confederates attacked the left wing at Seven Pines.

The health of the army was now of grave consideration. In Battery E a large number were sick, and several had died. An extensive hospital was established at Savage Station, but those who could be moved were transported to Northern hospitals. In clearing up the battlefields of Seven Pines and Fair Oaks, only shallow graves could be dug for the dead, owing to the large amount of water in the soil, which would fill them at the depth of a shovel, therefore the bodies were laid in rows, sometimes in heaps, and the earth thrown over them. Some did not have even such a burial as this, as those in Battery E, who went into the woods on the right, or the slashing in front of our redoubt, can testify. The sight then was sickening, many bodies lay as they fell, and struggling, died. Some were covered with earth, but generally only a few shovelfuls were thrown on their bodies, the arms and legs being exposed, and some were covered with sticks of wood from a pile there was in the woods. A large number of horses that had been killed were also carelessly buried, or left uncovered in the same manner. The air we breathed was putrid, and the water we drank seemed tainted with the foul drainings of the battle-field. Add to this the drying up of the swamps and the malarial atmosphere, and can it be wondered that a pestilence did not threaten the army, or prevail before? The statement that McClellan lost more men by disease than by the bullet, is not untrue, as can be shown by the roll of every company and regiment in the Peninsular Campaign. On the 20th of June the surgeon-general's report showed the number excused from duty on account of sickness to be thirteen thousand.

The horses of the battery were sheltered from the sun in the woods in rear of the intrenchments, where a little camp had been made for the baggage, and material not used in active operations. Among the fallen trees in front were the pickets of both armies, and beyond this, about five hundred yards distant, was heavy wood, in which the enemy awaited our attack. In this they seemed to grow impatient, and a conflict between the pickets was of daily occurrence. Sometimes a shell from a rifled cannon would burst over our heads, or a solid shot go screeching a long distance to the rear. The sharpshooters' bullets made it dangerous for our cannoneers to be seen above the breastworks, and we never left our guns day or night.

At an early hour on the morning of the 25th, Generals Hooker and Kearny tried to advance their picket line in front of Seven Pines, which brought on quite a severe engagement, lasting until afternoon. In this fight the Second Rhode Island, which had returned to Couch's division, took part, losing several men, besides having about twenty wounded, among whom were Capt. William B. Sears and Fred A. Arnold, members of this Society. Sickles's brigade, on the left of our battery, was engaged, and suffered considerable loss. In this battle, Oak Grove, or King's School House, as it has been called, the loss in killed and wounded amounted to about six hundred, and, although our object was accomplished, the engagement was given but little attention on account of more important movements of the army the succeeding ten days.

At last the clamors of our people for the army to move on, and the cry "On to Richmond!" were ended. The Confederate generals were tired of waiting for McClellan to make the attack, and decided to make it themselves. On the evening of June 25th General Jackson crossed to the rear of McCall's division, which was on the right of the Union line, thereby awakening McClellan from his dream. The next morning General Lee, who had been in command of the Confederate army since the battle of Fair Oaks, drove in the advance of our right wing, and, in the afternoon, made a furious attack at several points. The rebels, largely outnumbering McCall, dashed upon his line with great courage, but were everywhere driven back. The action lasted from three until nine o'clock, during which time the thunder of artillery was distinctly heard the whole length of the line, and the enemy in front of our division made a feint in order to divert the attention of McClellan from the real object or plan of their attack.

During the night McCall's division was withdrawn some five miles from Mechanicsville, the scene of the engagement, and concentrated with Porter's corps, near Gaines Mill and Cold Harbor. That morning, June 27th, I accompanied an officer around to the right, to corps headquarters at Fair Oaks, thence to the headquarters of the commander-in-chief, at Savage Station, where I saw the entire baggage train, numbering, it seemed, thousands of teams. I had been at the same place two days previous, but the scene had so changed by the removal of headquarters' tents and the quartermaster's department, and the hurry with which army wagons were filled and started off with ammunition and provisions, gave rise to grave apprehensions in my mind, which I tried not to entertain. I had seen enough, however, to know that McClellan had planned for a retreat, and upon returning to the battery related it to Sergeant Williams, who immediately reported me to Captain Randolph as disheartening the men. I assured the captain that I had made no mention of my observation except to the sergeant, whom the captain found it necessary to restrain from going among the men telling what I had confidentially said to him.

At half-past two our lines were again attacked on the opposite side of the Chickahominy. The Confederates, who greatly outnumbered the division opposed to them, rushed on in great fury, but were repulsed and routed, the assailed became the assailants, and the rebels fled from the field in disorder. A lull then occurred in the battle, and we could tell by the sound of artillery that our troops held their ground; some thought they were advancing. About an hour before sunset the whole Confederate force, numbering full two to one of our own, made a simultaneous attack along the whole front, which was handsomely defended until twilight, when our line gave way at every point, falling back to the bluff bordering the Chickahominy, when French's and Meagher's brigades of reinforcements dashed upon the victorious rebels, checking the pursuit, and night ended the battle. Official reports of the number of men here engaged make it fifty-six thousand Confederates and thirty-six thousand Federals.

All the afternoon the Confederates showed themselves in front of Franklin, Sumner, and Heintzelman, with infantry and artillery, at times making quite an advance upon our pickets, so that it could not be told whether an attack was to be made or not. The rattle of drums was often heard as if there was a movement of large bodies of troops, the object in these feints being a diversion to prevent our sending reinforcements to the real point of attack. At night bands of music played Southern war tunes at different points in front of our line, and loud cheers of the exultant rebels gave us unpleasant reminders of their success.

During the night Porter and McCall crossed to the south side of the Chickahominy, destroying all the bridges on their retreat. Never before was the Army of the Potomac in a position to fight a battle. The whole force was within a space of three miles between the two wings. Between them and the rebel army ran the Chickahominy with its bridges torn up, and its high bluffs approachable only through a thick swamp. The front had changed to the rear, and behind our army lay the rebel capital stripped of its garrison, and only a few troops left to guard its gates. But McClellan had resolved upon a " change of base," and at daylight the retreat of the whole army was begun. In leaving our position the pickets were withdrawn, and a strong guard reserved within the intrenchments. Our guns were then run by hand into the woods, where the horses were attached, and, leaving everything that could not be conveniently carried, we quietly withdrew from before the enemy's pickets. We were soon joined by Thompson's and Beam's batteries, of our division, also several regiments of infantry, some of which had been marched to the right the day before, but arriving too late to take part in the battle, had just returned. Keyes's corps had taken the advance of the retreat, followed by Franklin and Porter. Heintzelman and Sumner fell back towards Savage Station, Battery E halting in battery at the edge of White Oak Swamp, where a line of battle was held until night. This was on the 29th. Heintzelman had orders to destroy all the stores at Savage Station that could not be carried away. The stores and provisions were piled up in great pyramids and set on fire. The ammunition was piled on a train, which with a full head of steam was fired and sent towards the Chickahominy. A great volume of smoke and flame, a terrific explosion, a dash from a burning bridge into a chasm, and all this that would have fallen into the enemy's hands was destroyed. About five o'clock the Confederates made a fierce attack upon Sumner's corps at Savage Station, in which they were repulsed and driven back with such a loss that they did not see fit to try it again.

Battery E spent the night on the road of retreat; not in bivouac, but slowly marching with long and frequent halts. A journey full of discouragement to say the least. Not whipped, but out-generaled. A flight from a lost battle. No one could tell where we were going, except it was thought that under the port-holes of the war vessels on the James we would find protection.

On the morning of June 30th we crossed White Oak Creek at Brackett's Ford. This was the third day our horses had been without food and water. In fording this stream the water reached their bodies, but they were not allowed to stop. We had no time, and they were only refreshed by the bath and the little they drank while moving across the stream. It soon became apparent that the enemy was close upon us, that a stand would be made, and that it would now be our turn to meet the rebellious host. At first we went into what seemed a very favorable position in an open field, where we waited for the approaching enemy considerable time, during which our men, as well as the infantry force about us, refreshed themselves with eating some low blackberries that were ripe and in abundance. Then we were ordered to another place, and, entering the woods by a narrow path, moved further to the left, and at last went into battery on a spot of not more than half an acre, covered with brush and surrounded with heavy woods. Captain Randolph soon resolved to get out of such a perilous situation, and, attaching the horses to the guns, again we moved away. Our lines by this time had fairly been established, extending nearly eight miles. The pop, pop, pop of our pickets denoted the advance of the Confederates. Battery E was in a most precarious position, and in order to get out our guide was obliged to lead us into the woods, through which we picked our way amidst the trees and between the enemy's pickets and our own. At noon the four rifled guns commanded respectively by Captain Randolph, Lieutenants Arnold, Sheldon, and Bucklyn, went to a position on the right in Slocum's division, where they soon became engaged. Heavy cannonading was heard in the direction in which McCall and Sumner were posted. It was here that Longstreet made the first attack, pressing forward his brigade in mass, but was unable to gain any ground. Foiled at this point the rebels dashed with great fury on the line held by Hooker and Kearny, who met them with undaunted determination. Before their terrific fire no troops could have stood, and they again fell back. Again and again the exasperated rebels dashed upon our lines at different points, determined to break them somewhere. In this manner the struggle continued till dark, when the pursuing enemy fell back, leaving us in possession of the field at nearly every point. In this battle the four rifled guns were engaged defending a road by which the enemy tried to advance, and they expended upwards of one hundred rounds of ammunition. After being separated from the first four guns, the fifth and sixth, in charge of Lieutenant Jastram moved to a small opening in the swamp in rear of Kearny's division. The position here was not adapted to the use of artillery, and the two guns were for a time exposed to a shower of over-reaching minies, ricochetting shot, and exploding shell. Late in the afternoon an aide to General Kearny dashed up and ordered Lieutenant Jastram to advance with the two guns. The cannoneers each took a round of canister from the caissons, which were left behind, and, entering the woods a short distance on the left, followed a narrow track that led to the field of battle. Here we saw strong evidence of the bloody work that was going on. In the road and on either side we met hundreds of men with blood streaming from wounds, who staggered to the rear, hoping to find a friend, or a stream in which to bathe their shattered limbs, or cool their thirst. Emerging from the woods we turned to the left, into a narrow field. Thompson's battery, which had been in action nearly an hour, we found still hotly engaged. Passing in rear of their guns at a gallop we took position closely on their left. At this point a struggle had just ended. In front of the guns and over the field, the bodies of the slain were thickly scattered, while a dense smoke that settled upon the earth, cut off all view beyond. Above the smoke, to a height of about ten degrees, glared the setting sun, which seemed to have been stained with the blood that was then being poured out upon the battlefield. Behind this dense cloud of smoke the Confederates reformed their brigades. Then, sweeping the ground before them with artillery, and giving a fierce yell, they burst upon us. Our cannoneers hurried in delivering their canister. The Thirty-sixth and One Hundred and First New York Infantry that supported our guns, met the enemy with a deadly fire. Then, in the midst of a prolonged and terrific shower of musket balls our infantry fell back between the guns. Thompson at this moment drew off his guns under cover of the thick smoke, and Jastram attempted the same movement. One of Thompson's guns was driven against a heavy gate-post, which caught between the wheels and the carriage body, as they left the field; and into this another was driven. Wedged together in this manner, the enemy swept down upon the cannoneers and took possession. At the same time this was being done the rebels poured in a second volley at close range. One of the wheel horses on the sixth piece fell across the pole, and in the agonies of death entangled the whole team. The infantry that we had been called upon to support had fallen back in rear of the battery, and the enemy in vast numbers swarmed toward us. In this position Lieutenant Jastram saw that nothing could be done but to abandon his gun, and, handing a spike to the corporal, it was driven into the vent, not, however, till the enemy was almost within a sabre's length. The other gun was saved by drawing it to where the horses were attached and retreating to the woods. With this the battle of Glendale, or Frazier's Farm was ended, and nowhere else on the whole line did the enemy force us back. It is said by those who

7

are familiar with the military life of General Lee, that he nowhere showed the skill that he displayed in forcing the attack on the retreating column at White Oak Swamp.

It was the gun to which I was attached that was lost, and, after leaving the battlefield, I returned to where the caissons had been left. Soon Lieutenant Jastram came up, and gave orders to remain there and he would find the rest of the battery. In this he did not succeed, and returned, but soon went off again. I was the only man with the two caissons except the drivers, and, leaving the field as it grew dark, entered a road on the line of retreat, where we were soon joined by Lieutenant Jastram, who had with him the rescued gun and three men, the others having wandered or skulked away, two or three falling into the hands of the enemy.

We followed Thompson's battery on the main road, which was filled with retreating troops, and, after going about a mile, we turned to the left into another road, and, by order of Captain Thompson, unhitched our horses and bivouacked for the night. It was then about nine o'clock P. M., and, finding a fire that had been kindled by those who passed before us, we made some coffee, and from our haversacks made a good meal. It was the first we had eaten except an occasional nibble on the march, since we left the intrenchments in front of Richmond. After this the few there were of us lay down and soon fell into a sound sleep, from which none were disturbed till daylight streamed above us. I was among the first to awaken, and to my astonishment found that Thompson's battery had gone and that we were entirely by ourselves. Lieutenant Jastram hurried the drivers in hitching up the horses and we quit the place and pushed forward as rapidly as possible to prevent being gobbled up by the enemy, who we knew would make an early advance. At first there was no one to be seen, then a few stragglers appeared from out of the woods where they had slept. These were steadily added to by sick and wounded soldiers, who hobbled on rude crutches they had made from the limbs of trees. It was a pitiful sight to see these poor men, who had hurried from their cots at Savage Station to avoid being left behind, delirious with the disease that the Chickahominy had spread over the army, while others, with swollen and undressed wounds, with stiffened and shattered limbs, crept along as best they could by each other's aid. With such as these we filled our gun carriage and two caissons with as many as could hold on, and in this way continued the retreat for about three miles. At this point we fell in with the rear guard of our army, and continued on with them until about eight o'clock, when we entered upon the slope of Malvern Hill, where we met Captain Randolph, who was overjoyed at seeing us again.

I shall never forget the scene as Battery E moved across the hill to take position for battle. It was the most picturesque of any military spectacle I ever witnessed. The hill was about one mile in length by half its breadth, the top of which was nearly free from trees, and the whole Union army was to be seen in full battle array. Batteries of artillery standing with drivers rein in hand, or moving across the field with cannoneers, some riding, some on foot. Regiment after regiment of infantry, with their glistening bayonets, standing " in place, rest," waiting for orders ; some marching to other positions, and

now and then a squadron of cavalry ready for
action. "Rush's" regiment of mounted lancers,
with their long pikes, near the point of which was
attached a quaint little red flag, was moving along in
column to a distant position. Not least of all,
were the siege train and the ammunition wagons;
also the large corps of ambulances, while here and
there dotting the plain were to be seen three-cornered
banners of various colors, indicating the headquarters of corps and division commanders, around which
general officers were dispatching their orders, and
aides with galloping steeds, dashed to and fro, adding
life to this intensely fascinating and beautiful scene.

Battery E, united again, except the gun lost in the
previous day's battle, moved easterly, halting on the
crest of the hill, near the centre of the line of battle.
The line was formed sickle like, extending around
the hill, both flanks resting on the James River, and
protected by the gunboats. The hill slopes gently
northward to the verge of a thick forest, while in
front and on the right, it fell somewhat abruptly into
a ravine. Between the hill and the woods, a distance
of perhaps one thousand yards, were level farm lands,

on which were patches of meadow and waving grain. In the centre of this field at our right were a dwelling and farm buildings, having more than an ordinary enterprising appearance. A part of Keyes's corps was attached to our own, that of Heintzelman, on the left of which, forming the left wing of the army, was Porter. Sumner united with us on the right, and next to him Franklin, and last on the extreme right was the balance of Keyes's corps. From the crest of the hill on which Battery E was situated, the whole field across which the enemy must pass to make an attack, was in the range of our guns. Thompson's battery on our right, and Bean's battery on our left, were formed at close intervals, it being only a few yards from gun to gun. Thus the whole plateau was surrounded with cannon, the heavy siege guns occupying a position in the rear. At the foot of the hill, sheltered by a growth of small trees and a slight ravine, was posted a strong force of riflemen, and behind our guns a strong support of infantry lay in close lines of twos and fours. In this manner the army was concentrated for resisting further advance of the enemy.

A sharp lookout was kept along the edge of the
woods beyond the plain until an occasional horseman
was seen, who seemed to be viewing our position.
Soon after this we saw a sudden puff of smoke from
a cannon that had been run up by hand, but still
masked by the wood, and a shot went screeching
over our heads. But before the smoke had cleared
away this was followed by another, and still another,
until there were five or six of them at different dis-
tances, their target appearing to be our battery.
Our cannoneers sprang quickly to their stations, and
the little duel was soon ended by the enemy being
put to flight. Various statements have been made
that there was no firing till about two o'clock P. M.
But I think I am not mistaken in placing this at half-
past ten in the morning. Half an hour after this a
banging commenced on our right, in front of Sumner, by the enemy, who had drawn a battery from
the woods, hidden by a barn, behind which and other
buildings on the premises they posted their guns.
Almost at the same time fifty guns were discharged
at them, setting fire to the buildings and compelling
them to withdraw. The rebels made some good

shots in their sudden attack, many of their shot barely clearing our heads, and two or three shells burst beautifully, as we were accustomed to say, over Thompson's battery. Sumner had taken up his headquarters under a stately elm in the yard of a farm house, and a shot that took away the jet and part of the tree-top caused a hasty evacuation of that locality.

Later, I should guess it was about noon, a battery of artillery came out of the woods at full gallop to about midway of the plain, and directly in front of Battery E. We could not help admiring the skillful execution of this movement, and the rapidity with which they went into battery and commenced firing. But sooner than they could fire a second round, our shell burst among them in such quantities that they became hidden in the smoke of bursting shell. The battery must have been shot to pieces, as we saw nothing of it again.

The battle was by this time fairly opened, and a fierce attack was made on us with artillery in various places along the line. A force of sharpshooters advanced under cover of the standing grain to within range of our cannon, and for a time was considerably annoying, but we succeeded in driving them off. The afternoon was wearing away when an artillery attack was made to the left, but not out of range of our battery. The enemy's guns were soon silenced, but we kept up a slow fire on the woods where we knew the rebel brigades were forming for an attack. At half-past five the enemy opened upon Couch and Porter with the whole strength of their artillery, and at the same time advanced with heavy columns to carry the hill, but the heavy fire of our guns forced them back. Brigade after brigade then came out from the woods and rushed with yells across the open plain, through storms of canister and shell, until within a few yards of our line, when our infantry poured in such volleys they were sent reeling to shelter. Again and again they repeated the attack, and were as often driven back, leaving the field strewn with their dead. At last our men held their fire till the advancing columns were nearly to the edge of the hill, when they poured in a single volley, and dashed upon the stunned and confused rebels with bayonets, driving them in disorder from the field. It was long after sunset when the battle ended, but not until after nine o'clock that our artillery ceased its fire.

During the engagement Battery E was well protected from the enemy's fire by the crest of the hill, which was almost to the muzzle of the guns. Behind us the hill sloped gradually for two hundred yards into a ravine, then rose to the level of the hill, where the siege guns were posted, firing at a high elevation. Behind us lay a brigade of infantry in support of the artillery, which rendered great assistance in bringing up ammunition from the wagons, that were sheltered some distance on the right. The casualties in the battery were slight, there being but one man killed and six wounded. The infantry that lay on the ground behind us lost severely, the Fortieth New York having twenty killed and about an equal number wounded, almost entirely by shot and shell that passed our battery. Captain Bean, of the Second New Jersey Battery, was instantly killed by a cannon ball while in the act of mounting his horse. The battery also lost several men, as did also the Second Regulars. During the engagement there were expended in Battery E more than three hundred rounds of ammunition to each of the rifled guns, and over five hundred rounds in the howitzer. The first shot fired on the Union side at the battle of Malvern Hill, was by Sergeant, afterwards Captain Lamb, from number one gun. The last sound of a cannon to close the seven days' battle was fired by Sergt. Joseph H. Milne, from the fifth piece, in charge of Lieutenant Jastram.

Before the firing ceased the army had commenced its retreat from Malvern Hill. Some refused to obey the order. General Martindale shed tears of shame. The brave and chivalrous Kearny said, in presence of many officers, "I, Philip Kearny, an old soldier, enter my solemn protest against this order for retreat. We ought, instead of retreating, to follow up the enemy and take Richmond, and, in full view of all responsibility of such a declaration, I say to you all, such an order can only be prompted by cowardice or treason."

At ten o'clock Battery E left its position and fell in with the rear guard. The retreat to Harrison's Landing was by a single road, a distance of seven miles. At midnight rain commenced falling, and the road at once became muddy. The large force in advance was frequently compelled to halt while some broken down or mired wagon, or piece of artillery was removed from the road; and in the rain and darkness the troops hurried on till the whole army was one confused mass, trudging despairingly, overcome by defeat and fatigue till about six o'clock A. M., when the last reached Harrison's Landing.

With this closed the last of the six days' battles, which history has recorded as seven. No one who was there on the morning of July 2d can ever forget the look of discouragement on every man's countenance. The rain poured down in torrents, the whole army was without shelter, no food could be had, nor fires kindled. The dead that had died in the ambulances during the night were removed and left by the roadside, and men grouped beneath the trees or quietly strolled about in search of some dry place where they might sit down for rest. By half-past nine the rain had ceased. Division and brigade commanders were busy in separating their commands, which were huddled together without order or division. Before this had been accomplished a battery of rebel artillery appeared on an elevation about a mile distant, and sent a dozen or more shells among us, creating for a time considerable confusion. Three of these shell exploded in our battery, but without serious injury. The first section, Lieutenant Arnold, started immediately with a force of infantry and cavalry, in pursuit of the audacious bombarders, and drove them away.

The next day, July 4, 1862, the Army of the Potomac began intrenching itself, with both flanks resting on the James River, under protection of our war vessels, and here I will leave my story of Battery E. I have not said all that should be said of the officers and men of this company, or of those noble hearted and patriotic young men who died of disease in the swamps of the Chickahominy, or those captured and taken to Southern prisons, or who fell in battle, or by the roadside, as the army was pushed back in its retreat. I may have done an injustice to a company of men which was second to no other that went from our State, by crowding into this short space a year of service, within which was a campaign of blunders ever to be remembered.

8

.

www.ingramcontent.com/pod-product-compliance
Lightning Source LLC
Chambersburg PA
CBHW021955190326
41519CB00009B/1274